EMP Survival Guide:

How To Survive An Electromagnetic Pulse Attack and Prepare Yourself For Living After The Power Grid Goes Down

Disclaimer: All photos used in this book, including the cover photo were made available under a Attribution-Non Commercial-Share Alike 2.0 Generic and sourced from Flickr

Table of content:

- Introduction: What is it? And how was it Discovered? 4
- **Chapter 1: Potential Weapon Applications of EMP 5**
  - Electro Magnetic Pulse Generator 5
  - Hemp 6
  - Nemp 7
  - E-Bombs 8
  - High Power Microwave Weapons 9
- **Chapter 2: Ground Zero of an EMP 10**
  - Strange Silence and Lights Out 10
  - Spoiled Food 12
  - Distress in the Hospital 13
  - Social Chaos 14
- **Chapter 3: Is EMP a Real Threat? And By Whom? 15**
  - Surprise Attack by Russia 16
  - Chinese Retaliatory Strike 18
  - North Korea Makes Good on its Threats 20
  - Iran Drops Secret Bomb 22
  - Terrorists Deploy Suitcase EMP 23
- **Chapter 4: What Defense is there from EMP? 24**
  - Metallic Shielding 25
  - Tailored Hardening 26
  - Preparing the Private Sector 27
  - Detecting and Knocking out An EMP Device 28
- **Chapter 5: Important Food and Medical Supplies 29**
  - First Aid Kit 30
  - Garlic 31
  - Gauze 32
  - Canned Goods 33
  - Save Some Beef Jerky 34
- **Conclusion: Surviving an Electro Magnetic Pulse 35**

Introduction: What is it? And how was it Discovered?

The acronym "EMP" stands for "electromagnetic pulse". Traces of this electromagnetic pulse were first discovered with the detonation of high yield conventional weapons. But it wasn't until the first Atomic Bomb test in July of 1945 that the effects of a large EMP burst was seriously considered. Those who worked on the Manhattan Project for the then top secret U.S. atomic weapons program were duly informed of this possibility.

Further testing of high altitude nuclear explosions in the 1950's then confirmed what researchers had long suspected all along. And it was determined that the electromagnetic pulse of a strong enough nuclear grade weapon could potentially knock out the power grid of a substantially large region, potentially even an entire nation.

But not only would an EMP knock out the standard power structure it would also fry just about any electronic device within its reach! This means that in the immediate aftermath your car wouldn't start, your cell phone won't work, and all other electronic devices and appliances in your home would be completely fried and useless. This is how a frightening new possibility of warfare known as "EMP" had been discovered.

Chapter 1: Potential Weapon Applications of EMP

As damaging as EMP weapons can be to civil infrastructure, since they are generally non-lethal in nature, their application can be very tempting to state actors. These weapons allow for a significant disruption of an enemy's capabilities without the moral baggage of being responsible for massive civilian casualties. But a truly nefarious nation could potentially use these EMP weapons in a variety of ways to *maximize casualties and destruction*. This chapter explores all of the possible applications of weapons grade EMP.

Electro Magnetic Pulse Generator

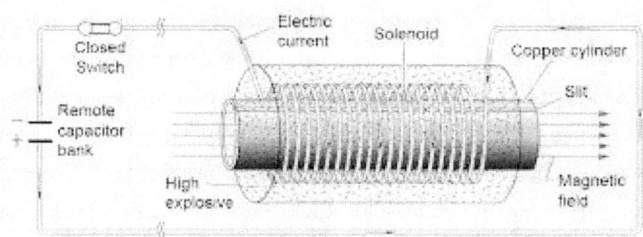

Electro magnetic pulse generators are still in development but if completed these highly focused, powerful weapons could potentially utilize microwave energy to create a massive and narrowly focused pulse that would be devastating both on the battle field and in the civilian theatre.

Hemp

Tested with devastating success, "HEMP" refers to an instance of detonating a powerful nuclear weapon high in the atmosphere, in order to send out intense electro magnetic fields over a wide area of land below. The electromagnetic pulse created from just one instance of HEMP would be enough to fry almost all of the electronics in a large area. The wave of this pulse could potentially take out radio towers, and most power lines and cables. A HEMP weapon would also interact directly with the Earth's magnetic field, and utilize gamma rays to increase the power of the electromagnetic pulse.

Nemp

Nuclear electro magnetic pulse weapons are the standard weapons application of EMP that occurs as a direct result of a regular nuclear bomb blast. This is the general side effect of any nuclear blast but is particularly classified as such when it is intentionally use in this fashion.

E-Bombs

So called "E-Bombs" are highly efficient non-nuclear bombs that can unleash an intense magnetic field. These highly specialized weapons could be potentially produced so small that they could be fitted inside a suitcase. It is said that these weapons may be highly effective on the battlefield against specific military targets but would not pose that much of a threat to larger civilian infrastructure.

High Power Microwave Weapons

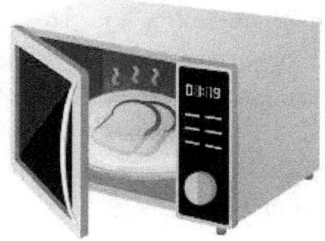

Described by the wavelength of their electromagnetic power, microwaves are used for a wide variety of applications from everything from radar to the microwave oven in your office that pops your popcorn. But HPM's or "high power microwaves" are on a whole other scale when it comes to the level of energy that is being produced.

In recent years many strides have been made to focus these HPM's into an EMP inducing weapon. By using something called a "flux compressor" these High Power Microwave Weapons can produce an incredibly formidable electromagnetic pulse that could wipe out any electrical equipment in its path. Even traditionally hardened and shielded devices are still affected by the penetrating wavelengths of High Power Microwave Weapons.

Remember when your mom told you not to stand in front of the microwave? Well there is a reason for that, because microwaves are able to permeate just about every surface, but unlike your relatively harmless microwave oven, a powerful HPM could wreak untold havoc.

Chapter 2: Ground Zero of an EMP

The ground zero of an EMP attack is certainly quite different from the ground zero of a nuclear or even conventional explosion. You won't find buildings shelled to their foundation or dead bodies smoldering in the street. The effects of an EMP are invisible and they only affect machinery, *not biological material* such as human beings consist of! In this chapter we will guide you through just what you might expect to find at the epicenter; *ground zero of an EMP attack!*

Strange Silence and Lights Out

The calm quiet of the aftermath of an EMP burst is one of the first standout features of such an attack. Imagine sitting down in your living room during a typical afternoon with TV blaring the news, dishwasher whirring through the latest round of dishes, your kids blaring music down the hall, and your next door neighbor cutting grass with the roar of a lawn mower. But in a split second, suddenly everything stops, no more TV, no more dishwasher, no more music, and even the lawnmower; inexplicably go quiet.

At first you think there must have been a local power outage; a simple enough explanation right? You instinctively reach for your phone to dial up your local power company, but you find the screen blank and lifeless.

You stare in disbelief thinking, "Didn't I just charge my phone? What is going on here?" Unable to call out for help you decide to take a step outdoors. But as you step outside and greet your neighbor whose *gas powered* lawn mower strangely conked out right when your power went out, the two of you turn to see another neighbor's car slowly skidding to a halt at the end of the street.

Another odd coincidence but you shrug it off as inopportune car trouble. You then see another neighbor futilely attempting to start his own car, putting in his key and cranking the ignition, but nothing but silence. Soon enough you would find that this silence in the aftermath of an EMP attack is permeating your entire neighborhood and the entirety of the ground zero of the electro magnetic pulse that has been unleashed. If the EMP burst occurred in the middle of the night the effect would be even more dramatic because all of your lighting would go out.

Even more disturbing, you would find that the trusty flashlight app on your phone is useless because all of your cell phones are fried! And even if you manage to fish out an old fashioned standalone "flashlight" (remember those?), you would be greatly distressed to find that even this simple instrument of illumination wouldn't work either! Your classic flashlight and the additional pair of batteries you stashed away, would be completely useless.

This is just how pervasive the electromagnetic pulse is, although you as a biological human being do not feel the pulse, in a split second it has flashed through every single piece of equipment you own. Sitting in the dark, you and your family will quickly realize that the only source of lighting available to you would be an old fashioned candle. In the aftermath of an EMP many will be wandering through their blackened neighborhoods with makeshift lanterns looking like they came straight out of the 1800's. This is the strange new world that is the ground zero of an EMP.

Spoiled Food

Besides the loss of our electronic devices, communication, and being able to travel freely down the road, the next biggest struggle will come from our refrigerator. Being left without power for days on end will obviously cause a massive spoilage of food. Knowing as much—in the immediate aftermath of an EMP attack—you should rank all of your food from the "most perishable" to the "least perishable".

You should then make it a priority to eat the most perishable items first. This means that for all of your food that you know will not last more than a couple of days unrefrigerated, you should eat as much of it as possible, you could even be a bit altruistic and share some of your food with your neighbors.

It's going to go bad anyway so you might as well create some good rapport with your neighbors. The least perishable food such as crackers, canned goods, and dried pasta, should be saved for the long haul, since this is food that you can rely upon in the intervening weeks without fear of it spoiling.

Distress in the Hospital

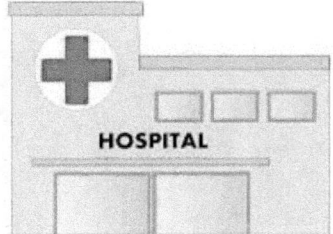

Even though an EMP blast would not directly kill anyone on the ground, the after affects could lead to significant casualties. And the most vulnerable of our population—those being cared for in our hospitals—would face the most dire of consequences. If an EMP successfully knocks out all power to the local grids, even generators wouldn't work and hospitals would truly be in the dark.

Patients in critical condition hooked up to life support would be the first to perish, flat lining as the machines that help keep them breathing go offline. After these patients perish, the next in line are those that are not necessarily hooked up to life support apparatus but who need certain life giving therapies and treatments that the power blackout would deprive them of.

After a few days these patients would die as well. One of the most disturbing aspects of an EMP is the high level of dead that it would leave in one of the most vulnerable segments of our population; those suffering from medical conditions. As you can see, such a dastardly attack would indeed bring about massive incidents of distress in our hospital system.

Social Chaos

Complete social breakdown would be the biggest fear of those trying to keep the peace in the aftermath of an EMP. With food spoiled, neighborhoods completely blacked out, and with no transportation to flee, citizens will feel cornered, and the propensity of civil unrest will be extraordinarily great.

How do you control a population that is on the brink of starvation, suffering a wide variety of medical illnesses, and with no solution in sight have lost all faith in government? It would be a difficult task for any leader to rein in such social chaos. And in such a tragic situation, a nation's own members could become the greatest threat of them all.

Chapter 3: Is EMP a Real Threat? And By Whom?

North Korea has been threatening to "incinerate" the United States for several years now. And now that the U.S. has a President that seems to fight fire *with fire*, even going so far as to state to the United Nations that if the U.S. is provoked, North Korea will be "destroyed", the threat from N.K. seems to loom largest. But North Korea is certainly not the only country that could surprise Western Civilization by turning the lights off with an EMP blast. There are several actors that could carry out such an attack. In this chapter we will outline some of the more plausible of these nightmare scenarios.

Surprise Attack by Russia

Many have joked and poked fun at those who bring up any concern in regard to tensions between the United States and the former Soviet Union. But although the Cold War is long over, and we live in a very different world, the fact still remains that the United States and Russia are the two most powerful militaries on the planet and even if these two nations are the best of friends today (contrary to what most policy wonks believe) it wouldn't take much for this relationship to deteriorate.

And having that said, Russia—of all nations—is rumored to have perhaps the most advanced EMP program in existence. So if push came to shove, there can be no doubt that the Kremlin has contingency plans of its own to use EMP against the United States. If Russia believed that a nuclear showdown with the United States was imminent for example, the prevailing theory is that Russian leadership would attempt to bypass the consequence of "mutually assured destruction" by launching such a devastating EMP that the U.S. would be knocked out before it could launch a single nuclear missile.

For such a thing to occur, the Russians would indeed have to have a powerful *and very precise* EMP weapon, and they would have to have a great deal of luck to create such a perfect electro magnetic storm that the U.S. is blacked out from coast to coast (they would also have to hope that U.S. nuclear subs wouldn't be able to reach them before being disabled by the pulse). Such unlucky odds for America may seem far fetched, but even the slightest chance of such a negative result needs to be taken into account.

Chinese Retaliatory Strike

Even in the best of times the United States and China have a rather tenuous relationship. There are quite a few ways that the U.S. and China can rub each other the wrong way and spark a devastating military conflict. For one thing—and as you can see is a common theme in this book—China is loosely allied with North Korea, and in light of the recent threats being leveled by North Korea to the United States, a conflict could erupt that pulls China right along with it.

Just how would such a nightmare scenario occur? Quite easily with the current stance that leader Kim Jong Un of Korea has been taking. Mr. Kim has been persistently pushing the limits of what the United States and the rest of the world can take. The world community is quickly discovering that if they tell Mr. Kim not to do something, in petulant defiance he will do it anyway. He was told not to launch missiles into the Sea of Japan, so he launches two missiles in rapid succession the next day.

And then after making threatening insinuations about attacking the U.S. military outpost of Guam in the middle of the Pacific Ocean, Mr. Kim was sternly warned to "not even think about it". So what does Mr. Kim do? The next day he informs the world media that he won't directly hit Guam—oh no, that would be crazy—but he would like to just set off a huge nuclear bomb right off the coastline instead!

The boyish leader of North Korea is literally testing the waters, pushing and pushing, inch by inch, to see just how far he can take things.

If North Korea did drop a nuclear bomb off the shore of Guam, while it may not lead to direct casualties, obviously the United States could not (without completely losing face and all credibility) stand by and let North Korea get away with blowing up nuclear bombs just outside their harbor! If the U.S. allowed this they would have to allow anything. Then again—if the U.S. does respond to this severe provocation, the crafty North Korean leader could rightfully proclaim that their nuclear test off the shores of Guam didn't technically hurt anyone.

Most would see the flaws in this logic and call this nonsense out for what it was, but push come to shove, China may decide to back its traditional ally. In this scenario, North Korea pushes its luck, takes things way to far, bombs the coast of Guam and forces the U.S. to strike North Korea. China alarmed that its neighbor is being incinerated, and fearing what might happen next, takes the initiative and drops a massive EMP over the U.S. in order to freeze the U.S. assault in its tracks. Let's hope none of this nightmare scenario ever occurs.

North Korea Makes Good on its Threats

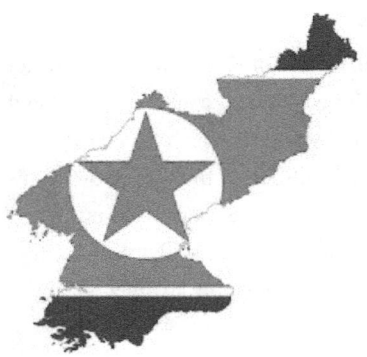

No one quite knows what Kim Jong Un (or as some have dubbed him; the "rocket man") is up to when it comes to his ambitious nuclear missile program. Many have made the claim that the North Korean leader only wishes to have a nuclear deterrent as a safeguard and collateral against any future U.S. attempt at regime change. This argument contends that while non-nuclear countries such as Iraq and Libya have faced grueling regime change either directly or tacitly backed by the United States, nations with nuclear weapons remain safe from such intrusions.

Directly feeding into this belief more than anything else is the glaring fact that the long time dictator of Libya, Momar Kaddafi actually voluntarily turned over his weapons of mass destruction, including the components of a nascent nuclear program with the promise from the Bush administration to never interfere with the Libyan government. But it only took a change of administration a few years later to have President Obama leading the charge to run Kaddafi out of power during the Arab Spring.

If Kaddafi had a nuclear bomb to play as his trump card, this probably never would have happened. It seems that Kim Jong Un has paid attention to this example of a head of state cooperating with nuclear disarmament only to perish, and taken the lesson to heart. This is why for many, it seems that Kim is primarily bluff and bluster, using his nuclear weapons to deter the United States, but would never be crazy enough to actually use them.

But then again, there are many who would point out that if all Kim sought was a nuclear deterrent, all he really needed was one nuclear bomb, just one nuclear weapon is usually enough to deter an outright land invasion of a country, yet Kim kept going after that *one bomb*, he kept going after *two bombs*, and he kept going at *20 bombs*. North Korea is now producing nuclear weapons—and ever more powerful and stronger grade nuclear weapons—at such an exponential rate, that some policy analysts are beginning to sound the alarm that a simple nuclear deterrent is not the only thing Kim is seeking.

For them it seems that Kim is attempting something else entirely. Kim Jong Un seems to have three goals in mind with his nuclear build up; either bully or destroy South Korea and unify the entire Korean peninsula under his reign, get revenge on Japan for the atrocities Japan committed against them during World War Two, and perhaps, just perhaps drop a massive EMP on the United States to make it unable to respond (at least in a timely manner) to North Koreas attacks on its neighbors.

According to this dreadful theory, North Korea would take out the U.S. electrical grid first with a powerful EMP, then bomb, and or flood troops into South Korea, while simultaneously decimating Japan with nuclear and or conventional weapons, hoping that the U.S. would be literally "powerless" to stop them. It is for a scary scenario like this that the U.S. needs to make sure that it can either prevent, or mostly withstand an EMP attack intact, so that North Korea would not be able to make good on its threats.

Iran Drops Secret Bomb

President Barack Obama made a deal with Iran to delay their nuclear program for 10 years, but many skeptical observers believe that Iran is probably continuing their development regardless. Coincidentally enough Iran already has long rage missiles courtesy of North Korea. The fact that North Korea has exported military hardware to Iran is what led President George Bush to make North Korea and Iran, part of what he termed the, "Axis of Evil".

On the surface, the two nations of North Korea and Iran do not have much in common. North Korea is a communist nation that eschews religion while Iran is a hardcore religious state. But despite their diametrically opposed belief systems, both nations have an equal sense of antagonism when it comes to the United States. And this is precisely why the American CIA under the Bush administration had such fear of these two nations actively collaborating with each other.

If Iran secretly produced a bomb, or if North Korea actually colluded with Iran enough to ship them an already constructed weapon, Iran could produce an unpleasant surprise in the form of an EMP burst over the middle of the United States. It would be with terrible irony that the so-called axis of evil would finally live up to all the hype and the fear mongering, by creating an electro magnetic variant of former President Bush's smoking gun.

Terrorists Deploy Suitcase EMP

So far in this chapter we have discussed the dangers of other nations subjecting the United States to a nationwide power failure through an EMP attack. Now lets explore the unpleasant possibility of terrorists using a much smaller, so-called "Suitcase EMP" to specifically knock out the power grid of a targeted city. In this scenario a small band of terrorists could unload a suitcase EMP in the middle of downtown New York creating a city-wide blackout.

Not only that, since the infrastructure was so thoroughly fried the terrorists are fully aware that it will take months for city officials to get the power back on. In such an attack, the black out itself would no doubt simply be the first step of the plan, and after the power grid is shut down, the terrorists would then move on to stage to which would be subjecting the disabled city to horrific terrorist attacks through conventional bombings, mass shootings, and even stabbings.

It's an awful thing to think about, but even a small suitcase EMP could lead to such horrible consequences. In order to exact such damage, it is believed that the total cost to finance such a mission—including the gathering of all technical components of the device—would cost less than a couple thousand dollars. A truly sobering statistic, and yet another reason analysts are staying up at night in the never ending struggle to keep the rest of us safe.

Chapter 4: What Defense is there from EMP?

The effects of an EMP are projected to be absolutely devastating to national infrastructure. So the question naturally arises; is there any way to protect against or prevent an EMP strike? As the protectors of a nation, institutions like the United States Pentagon have spent countless hours exploring every possible contingency plan to keep their country out of harms way. And when it comes to an EMP attack, many different contingencies for defense and preservation of the nation have been explored.

But it isn't just those on the national level that should carry the entire burden, each and every one of us as citizens should also be educated as much as possible on the means of our own survival. Having that said, this chapter uses a dual approach describing how the government as well as the individual citizen may be able to defend against the onslaught of an EMP.

Metallic Shielding

In order to use metallic shielding on electrical equipment, you will need to use a continuous piece of shielding such as copper or steel would provide. These metal shields usually don't completely cover the interior however, and will most likely consist of some exceedingly small holes for ventilation. Additional, auxiliary materials are usually used in order to compensate for this perceived gap in security. The shielding should be about half a millimeter thick in order to provide the best protection from the blast of an electromagnetic pulse.

Tailored Hardening

When it comes to tailor made hardening of electrical equipment in defense of a potential EMP attack, the first thing that needs to be considered is whether or not the system itself will be viable if hardening is achieved. With tailor hardening you are only encasing the most sensitive pieces of the electronics in metal cases. Being able to differentiate what part of an appliance or device needs hardening and what parts could do without is crucial in keeping an EMP defense within budget. But although this method is cheaper it has not proven to be quite as reliable as complete metallic shielding.

Preparing the Private Sector

Many have attempted to claim that there is not contingency plan when it comes to private sector infrastructure. This couldn't be further from the truth. Since civil infrastructure of the private sector could be severely damaged, measures have been taken to put into place powerful surge protectors that could not only withstand a bolt of lighting, but could also take on an electromagnetic pulse. This is a step in the right direction, but these surge protectors are by no means full proof and could easily be overwhelmed, but it is at least a start when it comes to preparing the private sector.

Detecting and Knocking out An EMP Device

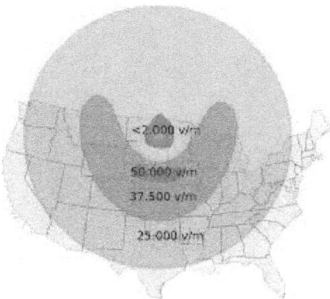

In recent years, primarily due to provocations from North Korea, the U.S. has stepped up its efforts to perfect methods of shooting down nuclear missiles and other offensive weapons, in mid flight. The conventional method is to use offensive missiles such as the "patriot" class interceptors to aim, shoot, and knock enemy missiles out of the sky. In addition to this, there are also projects in the works with laser technology to shoot down incoming objects.

The U.S. military has prototypes of ground laser batteries as well as continued research into space based laser platforms as was touted in Ronald Reagan's "Strategic Defense Initiative" program of the 1980's. And when it comes to something more subtle such as a suitcase EMP that terrorists are attempting to deploy in the middle of a city, man's best friend may be his dog after all.

There are several programs that have successfully trained the miraculous nose of the common dog to find and ferret out the components of what would make up an EMP device. So whether its through a patriot missile, a ground based laser, Ronald Reagan's SDI, or your dog's trusty sniffer, there are ways to detect and potentially knock out an EMP device.

Chapter 5: Important Food and Medical Supplies

Although there is not much the individual citizen can do if their power grid is knocked down—to turn the power back on—there are several other life saving aspects you can thoroughly prepare for. First and foremost on your list of preparation for an EMP attack would be adequate food and medical supplies.

Due to the nature of the crisis, these supplies would have to be completely nonperishable items that you could leave out in the item or packed away in boxes without any damage to the contents of the supplies. In this chapter we will explore some of the best of these nonperishable food and medical supplies to have on hand in the event of an EMP blast.

First Aid Kit

This is prepping 101 for just about any disaster, but yes, in the event of an EMP the traditional First Aid Kit would be a tremendous resource to have available. You should have a fully stocked kit with such medical supply staples such as cold packs, hydrogen peroxide, some Tylenol, and most importantly, needle, thread, and scissors. The latter items are of extreme import if you have to sow your own stitches. It may not always be pretty, but basic supplies like these will help you to survive the aftermath of an EMP blast.

Garlic

This herb is good to eat and heal your wounds at the same time! You probably recognize Garlic more for its use as a seasoning, but if you were to sprinkle a little bit of it in your wounds, it would work to greatly ease your pain and speed up the healing process. Garlic has special compounds that help to promote blood clotting and promote the formation of new platelets at the sight of the injury helping it to scab over and heal much faster than it would otherwise. So be sure to pack some garlic!

Gauze

The aftermath of an EMP can be a hazardous place, especially at night when it is difficult to see. These extra risk factors could lead to all kinds of injuries. This is why having a good roll of gauze on hand is critical to offset the danger. Immediately after you get injured wrap up the injury up with gauze and allow the area to pressurize in place so that it can heal. Gauze is an important medical supply to have.

Canned Goods

Some of the best food you could ever have stashed away in your cupboards are canned goods. These cans of food can last for decades. So yes, even though you may have laughed at your crazy uncle who kept a large supply of emergency canned goods big enough to fill a walk in closet, in our uncertain world such a practice is completely reasonable. In the face of an EMP attack when food might become incredibly scarce, such things don't seem quite so ridiculous! So yes, stock up on your canned goods!

Save Some Beef Jerky

Beef Jerky and other dried out meat are not just a good snack; they are a marvelous lifeline of protein preservation! Beyond beef jerky, just about any meat can be dried out with the right combination of salty brine, and through established drying techniques. If you don't know how to do it yourself, you can always purchase a nice supply of already dried beef and otherwise jerky! It's a great foodstuff to help see you through the worst of an EMP!

Conclusion: Surviving an Electro Magnetic Pulse

It is definitely a scary thing to wake up and find your entire city block out of power. It is horrifying then to find that you can't even start your car to flee the scene! But these are the startling realities of an EMP attack. In the chance that some rogue actor decides to implement such a devious strike against us, we have to be vigilant and we have to be prepared. I hope this book has left you just a little bit more informed, and more importantly reassured, of how you can survive an electromagnetic pulse. Thank you for reading!

www.ingramcontent.com/pod-product-compliance
Lightning Source LLC
Chambersburg PA
CBHW050252230526
45470CB00005B/2223